WORDS OF CHANGE
CLIMATE

WORDS
OF CHANGE

CLIMATE

POWERFUL VOICES,
INSPIRING IDEAS

CHRISTINA LIMPERT, PhD

SPRUCE BOOKS
A Sasquatch Books Imprint

FOR ELLIE AND EVERYONE
WHO BELIEVES IN THE
CAPACITY OF COLLECTIVE
CARE AND ACTS UP FOR
CULTURAL CHANGE AND
ECOLOGICAL STEWARDSHIP

We are in the beginning of a
mass extinction, and all you can
talk about is money and fairy tales
of eternal economic growth.
How dare you?!

—GRETA THUNBERG,

At UN Climate Action Summit
in New York, September 23, 2019

INTRODUCTION

As I write, a youth-led Black Lives Matter march moves through my neighborhood. The protest is over one thousand strong. An uncommonly warm spring breeze carries their righteous chants: "What do we want? Justice! When do we want it? Now! Say her name. Breonna Taylor! No justice, no peace. Black lives matter! Tell me what democracy looks like. This is what democracy looks like!" At this moment, we are experiencing an unprecedented collection of troubles. The death of George Floyd at the hands of the Minneapolis police triggered massive participation in a global movement for racial justice. We are also in the midst of a global pandemic that further highlights systemic social, racial, and economic inequality. At the same time, the devastating consequences of the climate crisis are crystal clear.

The fog of human discontent is thick. Across the globe, people who activate and occupy the real and virtual spaces of action have an urgent message: We have had enough. We have had enough of the brutal murders and unjust policing of Black, Indigenous, and people of color (BIPOC). We have had enough of policies that marginalize and put in harm's way members of the LGBTQIA+ community and people with intellectual and physical disabilities. We have had enough of corporations who pollute and pillage our land, air, water, habitats, and ecosystems. We have had enough of generations of politicians who place profit over people. Yet our rage is bookended by hope.

Unlike the past, today's climate movement is intersectional. That is, the movement for climate and environmental justice is also an inclusive movement that fights for racial, economic, gender, and social justice. In these pages, you will find inspiration from people—young and old, historic and contemporary—who are in the vanguard of the fight for climate and environmental justice. Today, climate activists like Greta Thunberg and Jamie Margolin walk in the footsteps of Rachel Carson who, decades ago, single-handedly raised the alarm about environmental threats. Artists like Jeaninne Kayembe and Xiuhtezcatl Martinez build on the work of Ansel Adams and Aldo Leopold, using art and storytelling as tools of empowerment and resistance. Youth like Katie Eder keep company with thought leaders like actress and entrepreneur Rosario Dawson and fashion designer Stella McCartney to challenge the status quo with bold, innovative ideas. Students like Jerome Foster II follow in the footsteps of Ayana Elizabeth Johnson, a marine biologist, to seek solutions through science. Indigenous activists like Eryn Wise and Tehontsiiohsta "Meadow" Cook carry generational wisdom and, much like Winona LaDuke and Oren R. Lyons Jr., apply traditional ecological knowledge to create a more just and sustainable existence.

Because this book is specifically aimed at youth, the majority of voices are from youth. I came of age in the 1970s and was ten years old when the first Earth Day was celebrated in the United States. In my generation we were often told that youth should be seen and not heard. To this day, in schools across North America, the stories and histories of youth and Black, Indigenous, and people of color are often papered over. But this global generation of climate activists will not be marginalized. Their urgent, cogent voices deserve to be amplified. Indeed, the words of change offered here remind us of what we share. The truth is, we face this struggle together, resolute in our efforts to seek reconciliation for a damaged planet.

Ours is a path of hope. Without collective and sustained action, the fabric of this movement will not be whole cloth. I invite you to join these activists, poets, scientists, students, potters, singer-songwriters, farmers, and everyday people as they prompt us to remember that we are not alone. Our collective ethic of care for people and the planet is a necessary artifact of our humanity. We are in this together for the long haul.

Lastly, I want to acknowledge that the land on which I live and work is the ancestral home of the Onondaga Nation of the Haudenosaunee Confederacy. The people of the longhouse are the oldest participatory democracy on the earth. Their model of democracy provided a framework for the American Constitution. Like the Haudenosaunee, many of us have a vision for a better future. The current confluence of troubles demands vigilant, collaborative action.

The time to act is now. Let us join together to fight for a more just, biodiverse, and habitable planet for all our relations.

Christina M. Limpert, PhD

We are in a world in which we no longer want a society of destruction but rather a society of change. . . . If we are doing that damage to Mother Earth, we are doing damage to ourselves.

—YANISBETH GONZÁLEZ

WE DON'T BECOME MOTHERS TO HAND OVER SOMETHING THAT IS DIMINISHING TO OUR CHILDREN.

—CHRISTIANA FIGUERES

"Conversations can't happen with an expectation of certain results, either that we will change someone else or that we won't be changed ourselves. They can only be done with a certain trust in ourselves, a trust in the others with whom we are talking, and a trust in the course of our shared life."

—TIM DeCHRISTOPHER

"We are each part of the planet's living systems, knitted together with almost 7.7 billion human beings and 1.8 million known species. We can feel the connections between us. We can feel the brokenness and the closing window to heal it. This earth, our home, is telling us that a better way of being must emerge, and fast."

—KATHARINE WILKINSON

WE MUST PUT LIFE BEFORE PROFIT.

–DAVID WICKER

For us, the nature is our supermarket, where we can collect our food, our water. It's our pharmacy, where we can collect our medicinal plants. It's our school, where we can learn better how to protect it and how it can give us back what we need.

—HINDOU OUMAROU IBRAHIM

"TALKING TO PEOPLE IN YOUR SOCIAL SPHERE ABOUT CLIMATE CHANGE IS ONE OF THE MOST CRITICAL THINGS YOU CAN DO ON A DAILY BASIS."

—CHERISE UDELL

"A society based on conquest
is not sustainable."

—WINONA LaDUKE

If we want to continue living on this planet, we have to solve the plastic pollution problem and we have to solve it within our generation. Humans have the incredible ability to innovate to survive at times when it matters. Now is one of those times.

—MIRANDA WANG

"I AM A PETITIONER BECAUSE THIS EARTH TAKES CARE OF US ALWAYS, NOW IT IS OUR RESPONSIBILITY TO DO THE SAME FOR HER."

—ESPERANZA SOLEDAD GARCIA

"YOU'VE GOT TO HOLD PEOPLE ACCOUNTABLE. YOU'VE GOT TO HOLD GOVERNMENT ACCOUNTABLE."

—PEGGY SHEPARD

It's the best feeling to be around kids who have the same interests as you and who are all united under one goal: protecting the planet. With the youth movement growing so quickly throughout the world, it's exciting to see our power and influence also grow as the attention is more and more focused on us. This gives me hope.

—ISABEL "SCOUT" PRONTO BRESLIN

A THING IS RIGHT WHEN IT TENDS TO PRESERVE THE INTEGRITY, STABILITY, AND BEAUTY OF THE BIOTIC COMMUNITY. IT IS WRONG WHEN IT TENDS OTHERWISE.

—ALDO LEOPOLD

Developing countries are said to be the most impacted from these changes due to the lack of resources, capital, technology and infrastructure. This will have a major impact on food production. If farmers are not trained or aware of ways to adapt to these changes, food production may be limited.

—NDONI MCUNU

The staggering biodiversity of this planet represents the diversity needed in this movement, a testament to the fact that there is a place for every single human to be involved and make a difference. . . . We're energizing each other for the long run as we act together for our future.

—MAIA WIKLER

THE LAND AND THE PLANTS ARE OUR RELATIVES.

—ANTHONY TAMEZ-POCHEL

"What we are saying is listen. Listen. That's where Indigenous people have something to add to this discussion because they talk about long-range thinking, they talk about seven generations, they talk about responsibility to the future. No discussion about gold or silver or money. Discussion about protecting water, protecting life. That's what this discussion is about now. "

—OREN R. LYONS JR.

The fight for truth—employing the principles of civil disobedience, nonviolence, and non-cooperation—is not just our right as free citizens of free societies. It is our duty as citizens of the earth.

—VANDANA SHIVA

"WE CAN'T JUST EXPECT STUDENTS TO RALLY AROUND A CAUSE THAT THEY DON'T RELATE TO;

"WE NEEDED TO
SHOW THEM IT
MATTERS TO THEIR
EVERYDAY LIVES."

—MERCEDES "SADIE" THOMPSON
AND CLAIRE WAYNER

"Our elders and frontline youth are the true celebrities. Never hold someone's social status above the needs of the people and remember when we step into these spaces we are all the same. We all need the fire, water, four winds and earth to exist, and for that purpose we have united."

—THOMAS TONATIUH LOPEZ JR.

I SEE A BEAUTIFUL
PLACE BUT IT'S
BITTERSWEET BECAUSE
IT'S BEING DESTROYED
AND ISN'T GOING TO
BE THERE BY THE
TIME I GROW UP.

—JAMIE MARGOLIN

THERE IS SOMETHING
INFINITELY HEALING IN THE
REPEATED REFRAINS OF
NATURE—THE ASSURANCE
THAT DAWN COMES AFTER
NIGHT, AND SPRING
AFTER WINTER.

—RACHEL CARSON

We need to be thinking about everything we build in the context of mitigating climate-change impact. It can't be just about aesthetics, but also about serving a purpose.

—KOTCHAKORN VORAAKHOM

The way that we talk about climate change is too compartmentalised, too siloed from the other crises we face. . . . The crisis of rising white supremacy, the various forms of nationalism and the fact that so many people are being forced from their homelands, and the war that is waged on our attention spans. These are intersecting and interconnecting crises and so the solutions have to be as well.

—NAOMI KLEIN

TODAY'S YOUTH DOES NOT HAVE THE LUXURY TO HAVE A CHILDHOOD ANYMORE.

—CIARA LONERGAN

"One day I will be an ancestor and I want my descendants to know that I used my voice so that they could have a future."

—AUTUMN PELTIER

"

As climate change and global warming present a real and immediate risk, it is a critical time [to] rethink our approach to adventure and how we justify the impact of our carbon footprint and the mark we leave on places.

—LIZZIE CARR

"

CLIMATE DENIALISM IS NOT AN OPTION.

—ALESSANDRO DAL BON

"You don't have to be a grown-up to do something. Children are allowed to help the environment. If they don't, they won't have a future. They won't have anything to go to school for."

—LILLY PLATT

What's most important, however, in doing this work is the opportunity to highlight the agency and leadership of communities that have been on the front line of environmental struggles over the last several decades, particularly Black women and Indigenous women. They kind of bear the brunt of much of the struggle, and helping to amplify their voices . . . has been key.

—INGRID WALDRON

"With climate change, what can we do right now? Changing who we are, our habits. Changing the foods we eat, how we travel. Taking direct action to make our community better. Planting trees, community action, community-based solutions that are already there. That's what we as individuals can do."

—KEVIN J. PATEL

ENVIRONMENTALISTS
COME IN ALL SHADES.
OUR MOVEMENT
IS AS DIVERSE AS
OUR PLANET.

—MARIAN MEIJA

When businesses and makers are committed to doing deep work—beyond plastic-free packaging or flashy "green" marketing campaigns—of questioning their role in impulsive consumption, ethical production, quality-over-quantity, systemic racism, and overall transparency, the 'real' work of sustainability happens.

—AMELIA WREDE DAVIS

"We're at a point in time where it's an emergency, and we're not seeing any action from our leaders. And if the people who are leading us aren't doing any leadership, then I will. . . . We're not going to stop because there's no point in having an education on a dead planet, and at this stage, that's what we're headed for. We're going to keep going and keep fighting because we're not going to let our future go away."

—HARRIET O'SHEA CARRE

WE NEED COURAGE, NOT HOPE. . . . COURAGE IS THE RESOLVE TO DO WELL WITHOUT THE ASSURANCE OF A HAPPY ENDING.

—KATE MARVEL

When I am daunted by the enormity of the world, and the work to be done to make it sustainable and equitable for everyone, I am reminded and encouraged to lean into my community. We are doing so much, yet come from people who have done *even more*.

—ERYN WISE

"

I fight to protect the wildlife and people in Alaska from drilling and oil and gas exploration because [it] will hurt the environment, the animals, and the caribou, [which is] really bad for the Native lifestyle.

—ISAIAH HORACE

"

"Fighting against systemic oppression . . . in our everyday lives can be a very daunting challenge, but I've found that doing this work alongside people that you grow to love makes organizing fun, despite the enormity of the task that we've inherited."

—ZANAGEE ARTIS

"Avoiding the climate and ecological crisis is the greatest obstacle of our generation. . . . The time of passive civilians whose only contribution is voting every few years and (best case scenario) occasionally sharing a sensationalist Instagram post of the Amazon burning is over. This is not enough."

—EYAL WEINTRAUB

"When I was a student activist, the common slogan was 'Never trust anybody over thirty,' which I personally abandoned some years ago myself as a philosophy. Now I actually had an article in the AARP [American Association of Retired Persons] bulletin calling for a gray-green alliance . . . summoning up the idealism of their youth and finding ways to work with their grandchildren to save the earth."

—DENIS HAYES

TEACHING PEOPLE ABOUT CLIMATE CHANGE IS THE FIRST STEP IN FIGHTING IT.

—IQBAL BADRUDDIN

> I HAVE SEEN RIVERS THAT WERE BROWN WITH SILT BECOME CLEAN-FLOWING AGAIN. . . . THE JOB IS HARDLY OVER, BUT IT NO LONGER SEEMS IMPOSSIBLE.

—WANGARI MAATHAI

"

A PANDEMIC DOESN'T
STOP A WATER
CRISIS AND IT SURE
DOESN'T STOP ME . . .
BECAUSE CLEAN WATER
IS ESSENTIAL.

—MARI COPENY

"

To change the world, we must recognize the intersections of our movements and struggles. True progress towards climate justice is impossible without solidarity with and support of movements for racial, economic, and gender justice.

—LENA GREENBERG

"I KNOW THIS IS WHAT I HAVE TO DO: GIVE A VOICE TO THE PEOPLE THAT HAVE BEEN SILENCED."

—HELENA GUALINGA

One person asking for inconvenient change is mostly inconvenient. Two, five, ten, one hundred people asking for inconvenient change are hard to ignore. The more you are, the harder it gets for people to justify a system that has no future. Power is not something that you either have or don't have. Power is something you either take or leave to others, and it grows once you share it.

—LUISA NEUBAUER

"WE'RE WORKING TO HEAL THAT INTERGENERATIONAL TRAUMA."

—TEHONTSIIOHSTA "MEADOW" COOK

I act for climate justice with seven generations behind me and for the seven generations ahead of me. I encourage all of us to stand back and listen to our elders, to give recognition to the original nations that took care and prayed on the land we live on today and to act more gentle and compassionate towards ourselves so that we are able to walk gently and compassionately on Mother Earth.

—MAYA LAZZARO

"For over 500 years we have been fighting to protect our land and people, and the bottom line is that our futures are at stake. . . . If we do not act now and make the switch to green and renewable energy sources, our grandchildren will not have the same access to clean air, land, and water as we do. My ancestors have roamed these lands since time immemorial, and within our teachings we are to live as one with the Earth, for she is our mother."

—NINA BERGLUND

Climate action is not just about reducing greenhouse gas emissions, it is also about creating green jobs in a post-COVID world, preserving our health and biodiversity, and protecting the poorest and most marginalized communities.

—NATHAN METENIER

THE BIGGEST
CHALLENGE WE FACE
IS SHIFTING HUMAN
CONSCIOUSNESS, NOT
SAVING THE PLANET.
THE PLANET DOESN'T
NEED SAVING. WE DO.

—XIUHTEZCATL MARTINEZ

"My advice to other activists is to get on Twitter. That's how I connected with other activists. And you don't have to wait for other people. On my first strike I was sitting alone for most of the day. Then people didn't know what climate strikes were. They didn't know Greta Thunberg. Now there are millions of us across the world striking and Greta speaks at the UN. It shows that you can change things."

—SAOI O'CONNOR

I'M AN AFRICAN. . . .
I AM A GIRL. I AM
A CHILD. I DESERVE
CLIMATE JUSTICE.

—LEAH NAMUGERWA

"We, the Indigenous youth, are standing here in solidarity with our brothers and sisters from the North. . . . And in the years ahead we'll continue fighting for the future that we deserve."

—MILITZA FLACO

"

IF WE WANT A FUTURE, WE'RE GONNA HAVE TO STAND UP FOR IT FOR OURSELVES.

—KATIE EDER

"

THERE IS NO CLIMATE JUSTICE WITHOUT RACIAL JUSTICE.

—ALEXANDRIA VILLASEÑOR

IT'S NO SECRET THAT THE BEAUTIFUL PLACES WE LOVE ARE IN DANGER.

—SIERRA ROBINSON

"Climate work is hard and heartbreaking as it is. . . . When you throw racism and bigotry in the mix, it becomes something near impossible. . . . So, to white people who care about maintaining a habitable planet, I need you to become actively anti-racist. I need you to understand that our racial inequality crisis is intertwined with our climate crisis."

—AYANA ELIZABETH JOHNSON

"My mum [late animal-rights activist Linda McCartney] used to say 'infiltrate from within.' [. . .] I choose to believe that I can make change, that I can show them a great example of a different way of doing business."

—STELLA McCARTNEY

THE CLIMATE CRISIS IS
AN IMMEDIATE THREAT
THAT RESPECTS
NO GEOGRAPHICAL
BORDERS NOR SOCIAL
BOUNDARIES.

—TIMOTHÉE GALVAIRE
AND TASSOS PAPACHRISTOU

"Injustice like this isn't new. What is new is that so much of your generation has woken up to the fact that the status quo needs fixing; that the old ways of doing things don't work; and that it doesn't matter how much money you make if everyone around you is hungry and sick; that our society and democracy only works when we think not just about ourselves but about each other."

—BARACK OBAMA

WE ARE ALL ONE BIG
FAMILY AND WE ARE
IN THIS TOGETHER. . . .
THIS IS A WAKE-UP
CALL TO EVERYONE.
EVERYTHING NEEDS
WATER TO LIVE.

—TOKATA IRON EYES

"So what can we do?
First we must do what we need to
do to process the trauma and grief
in ourselves and our communities
from the pain and damage that
has already been inflicted. . . .
Second, we have to vote. . . .
Third, we have to organize
to build solutions in our
own communities."

—CHARLIE JIANG

"It seems natural to me to want to protect our Mother Earth. I live in south Louisiana, where we lose a football field's worth of wetlands every hour, partially due to rising sea levels. Those wetlands soak up water and decrease the chances of flooding in our communities. My own home has flooded two times in the last two years."

—JAYDEN FOYTLIN

BE BOLD. WE'RE NOT GOING TO SOLVE THIS CRISIS UNLESS WE'RE BOLD.

—FELIQUAN CHARLEMAGNE

IF A COUNTRY DECIDES TO ACT ONLY IN ITS OWN SELF-INTEREST RATHER THAN ACTING COOPERATIVELY TO DEAL WITH THE CAUSES AND IMPACTS OF CLIMATE CHANGE, WE ALL STAND TO LOSE.

—ALEXIS McGIVERN

"It would be amazing if everyone could prioritize climate change, but the honest truth is that a lot of people don't have the privilege to do that when you're worried about taking care of your family or commuting back and forth between your work and making sure you have enough money to make ends meet."

—AYANNA LEE

"

THIS IS EVERY
MAN'S BATTLE . . .
EVERYONE NEEDS TO
DO THEIR PART IN
ORDER TO MITIGATE
THE EFFECTS OF
CLIMATE CHANGE.

—LINDSAY CODY

"

TOO MANY OF US STILL DON'T SEEM TO UNDERSTAND THE SCALE, SCOPE, AND SPEED OF THE CHANGES THAT THE CLIMATE CRISIS IS CAUSING.

—JEROME FOSTER II

"Nature needs us to speak up
and break the silence about
the violations against our
oceans and forests."

—KARLA STEPHAN

CLIMATE CHANGE IS OFTEN SEEN AS THIS FAR-FETCHED IDEA THAT'S GOING TO BE HAPPENING WAY AFTER US, BUT THE TRUTH ABOUT IT IS THAT IT'S HAPPENING RIGHT NOW.

—JUWARIA JAMA

"The climate crisis is the largest
threat to every single person
and living thing on this planet.
We must make sure that we include
everyone in our solutions because
everyone needs to be uplifted. This
movement led by Indigenous, front
line, and youth of color will win and
achieve a livable planet for all."

—NADIA NAZAR

DO WE SETTLE FOR THE WORLD AS IT IS, OR DO WE WORK FOR THE WORLD AS IT SHOULD BE?

—MICHELLE OBAMA

"If you have a strong connection to environmental issues, I think you should be influencing others to take the same steps as you. Those that are uninformed, inform them; those that want to do more but don't know how, teach them."

—BERTINE LAKJOHN

"

I BELIEVE IN THE
WISDOM AND TENACITY
OF YOUNG PEOPLE
TO LEAD THIS FIGHT
FOR CLIMATE JUSTICE.

—NADYA DUTCHIN

"

We fight for our Mother Earth because the fight for Mother Earth is the mother of all other fights. We are fighting for your lives. We are fighting for our lives. We are fighting for our sacred territory.

—ARTEMISA XAKRIABÁ

> **WE ARE TOTALLY RUNNING OUT OF TIME. . . . IT'S NOT ABOUT SOME PEOPLE TRYING TO DO THEIR BEST ANYMORE.**
>
> —ANUNA DE WEVER

"The trees don't know what color I am. The birds don't know what gender is. The flowers don't know how much money I have in my bank account. I think we can rely on nature to be the equalizer for us so we can shed that weight. The possibility is there for us."

—RUE MAPP

"Find time for nature
amidst development &
economic growth. If not,
we might end up having
nothing left around."

—JOHN PAUL JOSE

This is about your posterity, your planet, and your future. You can make a difference. You can help end the climate crisis. You can be the change you have been waiting for.

—MOHAMMAD AHMADI

"What is the purpose of this Youth Summit if two days from now you are letting fossil fuel corporations take the stage along with member nations and allowing them to influence climate policy when they are the ones who created this crisis?"

—SWETHA SASEEDHAR

IT TAKES ALL OF US RAISING OUR VOICES AS ONE TO TRY AND BE THAT TRIGGER THAT PUSHES THE SYSTEM TO ACTUALLY RESPOND TO THE CRISIS THAT'S AT HAND.

—JAMIE HENN

The whole world is, to me, very much 'alive'—all the little growing things, even the rocks. I can't look at a swell bit of grass and earth, for instance, without feeling the essential life—the things going on—within them. The same goes for a mountain, or a bit of the ocean, or a magnificent piece of old wood.

—ANSEL ADAMS

The climate space is a place
for you! Don't feel like
you have to be a scientist,
politician, or activist to
be in this movement.
The weight of the world is
heavy, and we need a lot
more hands to carry it.

—BRIANNA FRUEAN

"Beauty is a resource in and of itself.
Alaska must be allowed to be Alaska.
That is her greatest economy.
I hope the United States of America
is not so rich that she can afford to
let these wildernesses pass us by.
Or so poor she cannot afford
to keep them."

—MARGARET "MARDY" MURIE

Climate change is real. It is happening right now, it is the most urgent threat facing our entire species and we need to work collectively together and stop procrastinating.

—LEONARDO DiCAPRIO

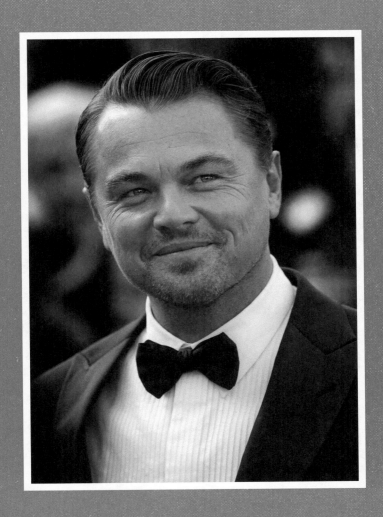

IT TAKES INCREDIBLE
COURAGE TO LIVE IN
A WORLD THAT YOU
KNOW IS ALREADY
LOSING SO MUCH TO
CLIMATE CHANGE.

—MAAYAN COHEN

"We must change not only our daily actions, but the mindset that let us become destructors to this planet. We must connect ourselves with the outside world and work together, because the ones causing the most harm are the ones not on the front lines."

—MADDY STEVENS

> WELL, I AM TELLING YOU, THERE IS HOPE. I HAVE SEEN IT. BUT IT DOES NOT COME FROM GOVERNMENTS OR CORPORATIONS. IT COMES FROM THE PEOPLE.
>
> —GRETA THUNBERG

"We don't have to keep hurting the planet. As we do our slow fashion push, the reality is [that] with fast fashion, if you're not paying for it, someone else is."

—ROSARIO DAWSON

WHEN YOU'RE NOT SPEAKING, YOU START TO TRULY LISTEN TO OTHERS AND LEARN WHERE OTHERS ARE COMING FROM.

—KALLAN BENSON

NOBODY SHOULD HAVE TO MOVE OUT OF THEIR NEIGHBORHOOD TO LIVE IN A BETTER ONE.

—MAJORA CARTER

"Climate policy is economic policy. . . . We have an opportunity to build better communities, with better health and mobility as well as greater access to wealth and resources for all people."

—RHIANA GUNN-WRIGHT

EVERY SMALL ACTION, REPEATED AND SCALED UP, HAS A HUGE IMPACT. START LITTLE, STRIVE FOR BIG.

—LOUKA DELAUTRE

"Humans must build a stronger spiritual reconnection to our mother, the Earth. An ethical relationship with creation, a natural bond. If that existed, among all nations, we would not be facing the catastrophic crisis we are in today."

—MAKAŚA LOOKING HORSE

"Environmental racism is a really big thing. The environmental movement is still predominantly white, how do we change that conversation? Having women of color leading is one way to do that."

—ISRA HIRSI

WITHOUT SOCIAL JUSTICE, WE CAN'T PROTECT THE NATURAL WORLD AND ACT ON CRISES SUCH AS CLIMATE AND ECOLOGICAL BREAKDOWN.

—HOLLY GILLIBRAND

"

JUST THE THOUGHT OF [MY FAMILY] NOT HAVING A PLACE TO LIVE BREAKS MY HEART. IF WE TAKE ACTION NOW ON CLIMATE CHANGE, THEY WON'T HAVE TO BECOME REFUGEES.

—KULSUM RIFA

"

"Climate change isn't just about
temperatures and weather; it's
about people. Our earth will be
here for millennia; it's up to us to
decide if humanity will be too."

—VIC BARRETT

[KIDS'] FUTURE IS SYNONYMOUS WITH *THE* FUTURE. AND THAT'S WHY THEY NEED TO FIGHT FOR IT.

—NAINA AGRAWAL-HARDIN

"Keep in mind that climate change makes every other issue you might care about worse. It impacts our security, it leads to more refugees, it impacts our health, and it worsens the situation for women, people of color, people living in poverty, and other marginalised groups."

—MONIKA SKADBORG

THIS COURT CASE IS ABOUT ESTABLISHING A CONSTITUTIONAL RIGHT TO A STABLE CLIMATE SYSTEM CAPABLE OF SUSTAINING HUMAN LIFE.

—KELSEY JULIANA

So much pain and suffering
for all the things living
on this earth—all made by
us. This is a fight that will
determine life and death for
so many; this is a fight that
is worth fighting for.

—HAVEN COLEMAN

"THE CLIMATE ISSUE
IS A LIFE ISSUE.
WE HAVE TO KEEP
FIGHTING UNTIL WE
SEE THE ACTION THAT
WE ARE DEMANDING.
ABOVE ALL, I WANT
PEOPLE TO STOP
DYING AS A RESULT OF
CLIMATE CHANGE."

—VANESSA NAKATE

In fact, *there are half as many wild animals on the planet as there were in 1970,* an awesome and mostly unnoticed silencing.

—BILL McKIBBEN

You are more powerful than you know. You can make changes to your life that seem impossible today. And you can do this, even if it means your greatest challenge right now is simply accepting the truth in "I do."

—NIKKI ERA

"OUR GOVERNMENT
CONTINUES MAKING
TODAY'S PROFIT
A PRIORITY OVER
PROTECTING OUR RIGHT
TO A SUSTAINABLE
FUTURE. MY GENERATION,
WITHOUT VOTING RIGHTS,
HASN'T BEEN FAIRLY
REPRESENTED IN MAKING
DECISIONS THAT WILL
AFFECT OUR LIVES."

—NATHAN BARING

We do not want to stop our ways of life. That's why we're here. We shouldn't have to tell people in charge that we want to survive. It should be our number-one right. We should not have to fight for this.

—QUANNAH CHASINGHORSE

BIOGRAPHIES

ANSEL ADAMS *(he/him)* was an American photographer, writer, and conservationist whose wide-ranging black-and-white natural landscape photos cemented the "wilderness idea" in the American psyche. Through Adams's persuasive eye we envision art photography as a means to inspire environmental preservation.

NAINA AGRAWAL-HARDIN *(she/her)* is a climate activist and organizer from Ann Arbor, Michigan. Like many in her generation, Naina found activism through social media. She works with several climate justice organizations including Zero Hour and the Sunrise Movement.

MOHAMMAD AHMADI *(he/him)* is a climate activist from Chicago. He is an outreach coordinator at Earth Uprising Chicago, a youth-led movement fighting the climate crisis.

ZANAGEE ARTIS *(he/him)* is a cofounder of Zero Hour. He was the logistics coordinator of the first youth climate march in Washington, DC. In addition to environmental issues, Artis advocates for gender and racial equality.

IQBAL BADRUDDIN *(he/him)* is a climate activist and founder of Fridays for Future Pakistan. Before leaving his job to work full-time on climate justice, Iqbal was a senior research associate at WaterAid Pakistan and a young professional officer at Leadership for Environment and Development (LEAD) Pakistan, a climate NGO.

NATHAN BARING *(he/him)* is a climate activist from Fairbanks, Alaska, and a plaintiff in *Juliana v. United States,* a lawsuit filed by the group Our Children's Trust against the federal government for violating the constitutional rights of young people by permitting activities that damage the climate. He is active in Alaska Youth for Environmental Action, Northern Alaska Environmental Center, and the Fairbanks Climate Action Coalition.

VIC BARRETT *(he/him)* is a speaker, organizer, and global climate justice activist, as well as a first-generation Honduran American from White Plains, New York. Vic is one of twenty youth plaintiffs in *Juliana v. United States*.

KALLAN BENSON *(she/her)* began her climate activism using silence as a tool for change. Her "silent strike" focuses on the act of listening. Now Benson is one of the organizers for the global movement, Fridays for Future USA, and co-organizer of the Outreach Working Group for Fridays for Future International. In 2019, Benson received Amnesty International's Ambassador of Conscience Award for her climate justice activism.

NINA BERGLUND *(she/her)* is a Northern Cheyenne and Oglala climate activist, speaker, and organizer who hails from Minnesota. Nina serves on Indigenous and Youth Councils at Earth Guardians, where she applies traditional ecological knowledge to support healing the earth on behalf of her ancestors, her community, and her homelands.

ISABEL "SCOUT" PRONTO BRESLIN *(she/her)* is an environmental activist who focuses on nature-based solutions for the climate. She founded youth-based Hudson Valley Wild, aimed at promoting wildlife conservation and biodiversity to address ecosystem damage.

LIZZIE CARR *(she/her)* used paddleboarding as a means to revitalize her health after a cancer diagnosis in 2013. When she found the waterways full of plastic debris, Lizzie woke up from what she calls her "environmental sleepwalk." Lizzie built Plastic Patrol, an app that crowdsources plastic pollution cleanup actions and provides data to reduce plastic pollution.

HARRIET O'SHEA CARRE *(she/her)* is one of the "Castlemaine Three," inspired by Greta Thunberg to initiate Australia's School Strike 4 Climate (SS4C) movement. She is an instrumental voice in the climate and human rights youth movement.

RACHEL CARSON *(she/her)* was an American biologist, science writer, and ecologist who penned the revolutionary book *Silent Spring,* which laid bare harmful agricultural practices and challenged governmental attitudes toward the natural world.

MAJORA CARTER *(she/her)* is an American urban revitalization strategist and award-winning advocate, organizer, speaker, and MacArthur Fellow. Originally from the South Bronx, she got her start as an environmental activist by focusing on the material consequences of air pollution on poor communities. In 2001, Carter founded Sustainable South Bronx to work on environmental advocacy and economic revitalization.

FELIQUAN CHARLEMAGNE *(he/him)* is the national creative director for Youth Climate Strike. He was born on the island of Saint Thomas and moved to Florida, where he saw the effect of climate change firsthand.

QUANNAH CHASINGHORSE *(she/her)* is a member of the Gwich'in and Lakota Sioux tribes and a speaker and activist working on behalf of the Indigenous peoples of Alaska to, among other things, advocate for a wilderness designation for the Arctic National Wildlife Refuge. Quannah serves on the Gwich'in Youth Council and the Steering Committee.

LINDSAY CODY *(she/her)* is a climate activist and organizer from upstate New York and the director of operations for Earth Guardians, a policy-oriented activist organization.

MAAYAN COHEN *(she/her)* is the director of partnerships and campaigns at the Alliance for Climate Education.

HAVEN COLEMAN *(she/her)* is an organizer, activist, and speaker hailing from Colorado, where she started as a climate striker with Fridays for Future. Haven is cofounder and national organizer of the US Youth Climate Strike and the founder and executive director of ARID, a creative agency for Gen Z.

TEHONTSIIOHSTA "MEADOW" COOK *(she/her),* whose name means "she who makes the land beautiful," is an Indigenous climate activist from Akwesasne Mohawk territory, of the Bear Clan. Her reservation in the North Country of New York State sits on a Superfund site, where a legacy of corporate environmental pollution is causing untold health effects on her community. She is the copresident and founder of the Three Sisters Sovereignty Project Youth Council.

MARI COPENY *(she/her)* is known as "Little Miss Flint" for her activism, advocacy, and fundraising in the ongoing water crisis of Flint, Michigan. When she was eight years old, Mari sent a letter to President Obama, which shone a national spotlight on the growing fight for clean water in underserved American cities.

ALESSANDRO DAL BON *(he/him)* is a youth climate activist and organizer with Fridays for Future from New York City.

AMELIA WREDE DAVIS *(she/her)* is a potter and interdisciplinary human based in Tacoma, Washington. She makes functional earthly vessels rooted in holistic sustainability and cultivating daily moments of pause.

ROSARIO DAWSON *(she/her)* is an American actress, producer, environmental activist, and co-owner with Abrima Erwiah of the African sustainable fashion brand Studio 189. Her civic engagement connects climate change with sustainable living, gentrification, and voter empowerment through a variety of environmentally focused campaigns.

TIM DeCHRISTOPHER *(he/him)* is a longtime climate justice advocate and organizer. He founded the Climate Disobedience Center, dedicated to growing climate activism through civil disobedience.

LOUKA DELAUTRE *(he/him)* was raised in a small village in France and studied at the University of Mannheim in Germany, where he founded the Earth Guardians crew. He is the regional director of Earth Guardians, Germany.

ANUNA DE WEVER *(she/her)* is an activist at the vanguard of the youth climate movement in Belgium. She started as a climate striker with Fridays for Future. In addition to her activism, Anuna holds an internship in the European Parliament with the Green Party.

LEONARDO DiCAPRIO *(he/him)* is an Academy Award-winning actor, environmentalist, and climate-change activist. In 1998, he founded the Leonardo DiCaprio Foundation to help spread awareness about the dangers of climate change.

NADYA DUTCHIN *(she/her)* is an environmentalist from Maryland. She is the coexecutive director of Power Shift Network.

KATIE EDER *(she/her)* is a youth leader, activist, and social entrepreneur who was named as one of *Forbes* "30 under 30" in law and policy. She is the cofounder and executive director of the Future Coalition, a network of youth organizers for social justice and change.

NIKKI ERA *(she/her)* is a singer-songwriter from New York. Her song "Monsters" speaks to issues of factory farming, environmental responsibility, and citizen involvement in greater change for our fellow earth inhabitants.

CHRISTIANA FIGUERES *(she/her)* is a public servant focused on policy reform and capacity building. She served as the executive secretary of the United Nations Framework Convention on Climate Change from 2010 to 2016, during which time she directed the negotiations that led to the historic Paris Agreement in 2015.

MILITZA FLACO *(she/her)* is a climate justice activist from Panama. She is a member of the Mesoamerican Alliance of Peoples and Forests.

JEROME FOSTER II *(he/him)* combines climate activism with social justice advocacy. He strikes with Fridays for Future in Washington, DC, and founded the get-out-the-vote initiative OneMillionofUs.

JAYDEN FOYTLIN *(she/her)* experienced the effects of climate change firsthand in her home parish in Cajun country in Louisiana. She is a plaintiff in *Juliana v. United States*.

BRIANNA FRUEAN *(she/her)* is an environmental advocate and climate change activist from Samoa. As a college student in New Zealand, she was a member of Pacific Climate Warriors, the youngest 350.org Samoa coordinator, and a TEDxAuckland speaker.

TIMOTHÉE GALVAIRE *(he/him)* and **TASSOS PAPACHRISTOU** *(he/him)* cofounded ECI Kerosene, which focuses on action at the policy level, specifically targeting intensive carbon-emitting industries such as airlines through the European Citizens' Initiative petition.

ESPERANZA SOLEDAD GARCIA *(she/her)* is a Xicana Indigenous activist, community organizer, and educator who fights for social justice through connection and collaboration. Esperanza works with various youth-led initiatives, including the Earth Guardians Youth Council.

HOLLY GILLIBRAND *(she/her)* is an environmentalist and activist, as well as a writer. As a youth ambassador for

Scotland: The Big Picture (TBP), she is a rewilding voice and an advocate for the restoration of wild nature and a healthier environment for all living things.

YANISBETH GONZÁLEZ *(she/her)* is an Indigenous leader from Panama representing the women in the Guna Yala territory.

LENA GREENBERG *(they/them)* is an American activist who works on issues of corporate accountability. She was a member of the SustainUS NYC Delegation to the UN Climate Summit.

HELENA GUALINGA *(she/her)* is an Indigenous activist from the Sarayaku community in Ecuador. Helena attends school in Finland and is part of the youth movement Polluters Out.

RHIANA GUNN-WRIGHT *(she/her)* grew up in Chicago, worked as an intern for Michelle Obama, and studied at Oxford University as a Rhodes scholar. She is a national policy expert and the leading architect of the Green New Deal. She is also the director of climate policy at the Roosevelt Institute.

DENIS HAYES *(he/him)* was the national organizer of the first Earth Day on April 22, 1970, an event that drew nearly twenty million people into the streets to demonstrate for change. As an environmental advocate, lawyer, and adjunct professor, Hayes has served on dozens of governing boards and founded the Earth Day Network. Hayes is the president of the Bullitt Foundation.

JAMIE HENN *(he/him)* is a cofounder and strategy and communications director at 350.org. He's also a contributing editor for ItsGettingHotinHere.org and an organizer for Step It Up.

ISRA HIRSI *(she/her)* is a Black Muslim environmental activist from the US whose goal is to build an intersectional climate movement. She is the cofounder and partnership director of the US Youth Climate Strike.

ISAIAH HORACE *(he/him)* is an Indigenous environmental justice activist from the Gwich'in Nation in Alaska.

HINDOU OUMAROU IBRAHIM *(she/her)* is an Indigenous activist from the Mbororo pastoral community of Chad; she is a geographer and an advocate for the inclusion of Indigenous peoples, along with their knowledge and traditions, in the global movement to fight climate change.

TOKATA IRON EYES *(she/her)* is a youth leader, water protector, and sustainability activist from the Standing Rock Sioux Nation. She is one of the youth leaders for the "ReZpect Our Water" campaign.

JUWARIA JAMA *(she/her)* is a student and coleader of the Minnesota Youth Climate Strikes.

CHARLIE JIANG *(he/him)* is a first-generation American-born Chinese climate campaigner at Greenpeace USA; he has served as an organizer for climate justice groups 350 DC and the Sunrise Movement.

AYANA ELIZABETH JOHNSON *(she/her)* is a marine biologist, policy advisor, and native of Brooklyn, New York. She is the founder and CEO of Ocean Collectiv and the nonprofit think tank Urban Ocean Lab. She is the coeditor of the anthology

All We Can Save: Truth, Courage, and Solutions for the Climate Crisis.

JOHN PAUL JOSE *(he/him)* is a climate activist and writer raising the public profile of the unchecked environmental crisis affecting India's forests and natural resources.

KELSEY JULIANA *(she/her)* hails from Eugene, Oregon, where she initiated her first climate-related lawsuit to address emissions reduction. She is the lead plaintiff in *Juliana v. United States*.

NAOMI KLEIN *(she/her)* is an award-winning Canadian author, activist, and filmmaker. Klein is a senior correspondent at the *Intercept* and the inaugural Gloria Steinem Endowed Chair in Media, Culture, and Feminist Studies at Rutgers University. She is the author of numerous books, including *The Shock Doctrine, This Changes Everything, No Is Not Enough,* and *On Fire.*

WINONA LaDUKE *(she/her)* is a member of the Anishinaabe Bear Clan and an activist, organizer, author, and a water protector who lives and farms in the Great Lakes region. An economist by training, Winona is an unflagging advocate for Indigenous rights, traditional Indigenous knowledge, restoration of foodways, rematriation of seeds, and building sustainable Indigenous economies. LaDuke established Winona's Hemp & Heritage Farm to create economic pathways for sustainable Indigenous agriculture.

BERTINE LAKJOHN *(she/her)* is a climate activist and youth leader from the Marshall Islands. A firsthand witness to the impact of climate change on her homeland, Bertine focuses her work on the threat of rising sea levels.

MAYA LAZZARO *(she/her)* is a Quechua and mixed European *warmi* (woman). She is a speaker and author who writes about water conservation and Indigenous knowledge and sovereignty, and serves on the Earth Guardians Youth Council.

AYANNA LEE *(she/her)* is a climate justice advocate from Milwaukee, Wisconsin, who served as the state lead and coexecutive director of the Youth Climate Action Team.

ALDO LEOPOLD *(he/him)* was a conservationist, forester, educator, writer, and sketch artist. His landmark book, *A Sand County Almanac*, published in 1949, rooted an ethic of care between people and nature in environmental thought.

CIARA LONERGAN *(she/her)* is a US-based environmental activist and logistics coordinator on the youth staff of Earth Uprising. Ciara is interested in meteorology and the effects of the climate crisis on weather.

MAKAŚA LOOKING HORSE *(she/her)* is a First Nations water activist from Mohawk Wolf Clan and Lakota, in Six Nations of the Grand River, Ontario, Canada.

THOMAS TONATIUH LOPEZ JR. *(he/him)* is the director of development for the International Indigenous Youth Council. Born and raised in Denver, Colorado, Thomas identifies as Mexican Otomi, Diné, Apache, and (Hunkapi) Lakota.

OREN R. LYONS JR. *(he/him)* is a Haudenosaunee Confederacy chief and faithkeeper of the Onondaga Nation. He is an internationally known environmental and Indigenous rights advocate from central New York State, where he lives on the sovereign Onondaga Nation, the central fire of the Haudenosaunee Confederacy.

WANGARI MAATHAI *(she/her)* is a Nobel Prize recipient, conservation activist, and women's rights champion. She founded the Green Belt Movement in Kenya. This environmental justice movement educates women on the economics of deforestation, and they've planted over fifty-one million trees.

RUE MAPP *(she/her)* is an outdoor advocacy expert and the cofounder and CEO of Outdoor Afro, a national nonprofit volunteer-based organization that uses outdoor recreation to reconnect African Americans with natural spaces, and build community and leadership in the process. Rue was a 2019 National Geographic Fellow and was the twenty-fourth Heinz Award recipient for the environment.

JAMIE MARGOLIN *(she/her)* is an activist, organizer, speaker, and cofounder of Zero Hour, a youth-led climate justice organization. A role model for her generation, Jamie is as comfortable carrying petitions around her neighborhood as she is testifying in front of the US Congress. She is the author of *Youth to Power*.

XIUHTEZCATL MARTINEZ *(he/him)* is an Indigenous activist, hip-hop artist, and author. He is the youth director of Earth Guardians. Martinez uses music as a tool for resistance and storytelling.

KATE MARVEL *(she/her)* is a climate scientist and an associate research scientist at NASA's Goddard Institute for Space Studies and at Columbia University's Department of Applied Physics and Applied Mathematics. Her research focuses on how human activities affect the climate.

STELLA McCARTNEY *(she/her)* uses her platform as a fashion designer to start conversations about the devastating environmental costs of fast fashion. In 2019, McCartney worked with activists from Extinction Rebellion to highlight the power of one of the world's most polluting industries as a way to address the climate crisis.

ALEXIS McGIVERN *(she/her)* is the education director at the Oxford Climate Society. She is a community educator whose work examines the impact of environmental pollution and change on marginalized communities.

BILL McKIBBEN *(he/him)* is a widely published author, educator, and environmental activist who, decades ago, sounded the alarm about global warming with his first book, *The End of Nature*. McKibben continues the fight through his work at 350.org.

NDONI MCUNU *(she/her)* is the founder of Black Women in Science, a nonprofit capacity-building organization that focuses on young Black women scientists and researchers. Her research focuses on food production and the effects of climate change on biodiversity and agriculture.

MARIAN MEIJA *(she/her)* is a first-generation college student and an artist and activist who was born and raised in the San Francisco Bay Area. Marian serves on Earth Guardian's Youth Council.

NATHAN METENIER *(he/him)* is the external relations officer for Youth and Environment Europe. He is from Grenoble, France, where he has witnessed the effects of climate change firsthand.

MARGARET "MARDY" MURIE *(she/her)*, known as the "Grandmother of Conservation," was a tireless advocate for environmental protection. Author, botanist, and conservationist, Murie's legacy includes the passage of the landmark Wilderness Act of 1964 and the creation of the Arctic National Wildlife Refuge.

VANESSA NAKATE *(she/her)* is a climate justice activist who formed the Rise Up Movement, Africa. She was motivated to act by the uncompromising heat and increasing temperatures, floods, and landslides in her homeland of Uganda.

LEAH NAMUGERWA *(she/her)* is a teen activist from Uganda and climate striker with Fridays for Future. Leah leads the fight for climate justice by marshaling others to fight deforestation and address damaging floods, landslides, and other environmental devastation wrecking havoc on crops, livestock, homes, and families on the African continent.

NADIA NAZAR *(she/her)* is an Indian American climate activist and artist who is also the cofounder, coexecutive director, and art director of Zero Hour.

LUISA NEUBAUER *(she/her)* is a student, activist, speaker, and climate striker with Fridays for Future in Germany.

BARACK OBAMA *(he/him)* served as the forty-fourth president of the United States from 2009 to 2017.

MICHELLE OBAMA *(she/her)* was the First Lady of the United States from 2009 to 2017.

SAOI O'CONNOR *(they/them)* is a climate justice activist, poet, organizer, and storyteller from Cork, Ireland, who is a vocal advocate for social and environmental justice.

KEVIN J. PATEL *(he/him)* is an environmental justice activist and organizer as well as the founder and executive director of One Up Action, a group that mobilizes youth to take direct action, such as planting trees.

AUTUMN PELTIER *(she/her)* is an internationally recognized water protector and clean water activist from Canada and the chief water commissioner for the Anishinabek Nation.

LILLY PLATT *(she/her)* is a global youth ambassador, speaker, and environmental champion from the Netherlands. Lilly spoke at the UN World Oceans Day 2020 about her innovative project, Lilly's Plastic Pickup.

KULSUM RIFA *(she/her)* came to the United States from Bangladesh when she was ten years old. As a child, she witnessed the destructive power of floods in her homeland. Today, Rifa is a leader with SustainUS and an activist dedicated to youth-led social and environmental justice.

SIERRA ROBINSON *(she/her)* is a farmer and permaculture teacher on Vancouver Island, British Columbia. She is the regional crew director of Earth Guardians and a member of Canada's Youth Council. She has engaged in regenerative practices since age eight.

SWETHA SASEEDHAR *(she/her)* was born in the South Indian state of Tamil Nadu and now lives in the United States, where she is involved with SustainUS.

PEGGY SHEPARD *(she/her)* is an environmental justice advocate and organizer from New York City and also the cofounder and executive director of WE ACT for Environmental Justice.

VANDANA SHIVA *(she/her)* is an Indian physicist, ecologist, author, advocate, and a social activist for seed sovereignty. A well-known ecofeminist, Dr. Shiva has written numerous books, including *Earth Democracy*.

MONIKA SKADBORG *(she/her)* is a Danish representative from the UNFCCC, where she serves as a liaison between youth and the government. She is an executive committee member for the European Students' Union.

KARLA STEPHAN *(she/her)* is an activist, organizer, and the national finance director for the US Youth Climate Strike.

MADDY STEVENS *(she/her)* is a climate activist who works with art as a way to highlight environmental issues facing us today.

ANTHONY TAMEZ-POCHEL *(he/him)* is a Chicago-born First Nations Cree, Sicangu Lakota, Black organizer, and the neighborhood services coordinator at City of Chicago. He is a politically active advocate and organizer for Indigenous youth through the Chi-Nations Youth Council. Among his causes are combating harmful race-based representations in sports and popular culture.

MERCEDES "SADIE" THOMPSON *(she/her)* and **CLAIRE WAYNER** *(she/her)* cofounded the youth-led Baltimore Beyond Plastic, which combines legislative and policy outreach to address plastic pollution. Mercedes is part of the sustainable food movement in San Francisco, California. Claire is active in environmental and climate justice organizations.

GRETA THUNBERG *(she/her)* single-handedly began the Skolstrejk för klimatet (School Strike for Climate), a worldwide movement that is now over eleven million strong, back in August 2018. A generation of youth activists credit Greta with motivating them to action.

CHERISE UDELL *(she/her)* is an organizer who founded Utah Moms for Clean Air. Her aim is to mobilize people to change laws and push for legislation that guarantees clean, healthy environments for all.

ALEXANDRIA VILLASEÑOR *(she/her)* is an American Latina climate activist, the cofounder of US Youth Climate Strike and founder of Youth Uprising. She is also on the advisory board for Evergreen Action.

KOTCHAKORN VORAAKHOM *(she/her)* is a landscape architect who addresses climate change through innovative urban design. She is the founder of the Porous City Network and lives in Bangkok, Thailand.

INGRID WALDRON *(she/her)* is a sociologist at Dalhousie University who studies environmental health and environmental racism in the Mi'kmaq and African Nova Scotian communities.

MIRANDA WANG *(she/her)* is an award-winning inventor, entrepreneur, and environmental advocate from British Columbia, Canada. Along with her partner and collaborator since high school, Jeanny Yao, Miranda is the cofounder and CEO of BioCellection, a firm that uses chemical technology to recycle plastics.

EYAL WEINTRAUB *(he/him)* is the cofounder of Jóvenes por el Clima Argentina (Youth for Climate Argentina).

DAVID WICKER *(he/him)* is a student activist, organizer, and climate striker with Fridays for Future. He is from Turin, Italy.

MAIA WIKLER *(she/her)* is a writer, researcher, and climate justice organizer for Indigenous and human rights. She is the media and communication coordinator for SustainUS. Wikler's writing can be found in *Colorado's Emerging Writers*.

KATHARINE WILKINSON *(she/her)* is a writer, environmentalist, and thought leader who runs Project Drawdown. She holds a doctorate in geography and the environment from University of Oxford, where she was a Rhodes scholar.

ERYN WISE *(she/they)* was born in Albuquerque, New Mexico. They are a two-spirit word weaver and member of the Jicarilla Apache Nation and Laguna Pueblo. Wise is the communications and digital director of Seeding Sovereignty and is co-organizing their Indigenous Impact Rapid Response Initiative. They were also a Standing Rock protest coleader.

ARTEMISA XAKRIABÁ *(she/her)* is an environmental justice activist of the Xakriabá people in Brazil.

PHOTO CREDITS

P. 126: Rosario Dawson: © Shutterstock.com/A Katz. ID 399808315.

P. 132: Isra Hirsi: © Shutterstock.com/Mike Jett. ID 1341075278.

P. 135: Vic Barrett: © Shutterstock.com/Lev Radin. ID 1510072715.

P. 137: Monika Skadborg: Reproduced by permission of Monika Skadborg.

P. 139: Kelsey Juliana: © Reuters/Kevin LaMarque. Stock.Adobe.com. ID 290575417.

P. 143: Bill McKibben: Reproduced by permission of Nancie Battaglia. All rights reserved.

P. 145: Nikki Era: Reproduced by permission of Nikki Era.

ACKNOWLEDGMENTS

First and foremost, I offer a resounding heartfelt thank-you to Sharyn Rosart, publisher extraordinaire, who walked me through this process with patient enthusiasm and expertise. I appreciate your visionary guidance and keen collaboration on this project.

Thanks, too, to everyone at Spruce Books for believing, as I do, that tweens, teens, and young adults are capable, committed agents of change. Thank you to Jill Saginario, production editor; Carrie Wicks, copyeditor; Alicia Terry, designer; and Nikki Sprinkle, Molly Woolbright, and Alison Sheridan in marketing and publicity who animated this book and lifted me up from behind the scenes. Likewise, I am grateful to Jenny Abrami, sales director, and the Penguin Random House sales reps, who work diligently to place books in the hands of booksellers.

Shout-out to the mighty collection of kids from our seventh grade English class at Manlius Pebble Hill School, who, after hearing about the Deepwater Horizon industrial disaster of 2010, jumped into action to organize a movie night fundraiser to help wildlife affected by the oil spill. You are young adults now, but because of you I know that even the smallest actions of everyday people sustain the greater good.

Lastly, big love to Quinn, Alexandre, Peter, and Penelope.

Printed in the United States of America

SPRUCE BOOKS with colophon is a registered trademark of Penguin Random House LLC

25 24 23 22 21   9 8 7 6 5 4 3 2 1

Editor: Sharyn Rosart
Production editor: Jill Saginario
Designer: Alicia Terry

Library of Congress Cataloging-in-Publication Data is available.

FOR A FULL LIST OF CITATIONS FOR EACH QUOTE APPEARING IN THIS BOOK, PLEASE VISIT CMLIMPERT.WEEBLY.COM.

ISBN: 978-1-63217-378-2

Spruce Books, a Sasquatch Books Imprint
1904 Third Avenue, Suite 710
Seattle, WA 98101

SasquatchBooks.com

LOOK FOR OTHER TITLES IN THE WORDS OF CHANGE SERIES

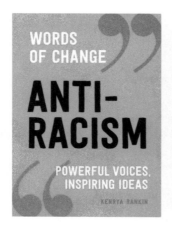

SPRUCE BOOKS

A Sasquatch Books Imprint